le Shuttle

THE OFFICIAL

CHANNEL TUNNEL

FACTFILE

D1535874

BⒺXTREE

First published in Great Britain in 1994 by Boxtree Limited
Broadwall House, 21 Broadwall, London SE1 9PL

ISBN 1 85283 357 2

Designed by Nigel Soper, Millions Design, London
Colour reproduction by Pressplan Reprographics Limited
Printed and bound in Great Britain by Cambus Litho Ltd,
East Kilbride, Scotland
Printed on Fineblade Smooth supplied by Denmaur Papers Ltd

A CIP catalogue entry for this book is available from the British Library

Words and phrases printed in **bold type** are explained in the Glossary.

CONTENTS

A TRIP THROUGH THE TUNNEL

For the first time since the last Ice Age, which ended about 12,000 years ago, it is possible to make a land crossing from Britain to France. At the end of the Ice Age, the melting of the ice caps raised sea levels, and Britain became separated from the rest of Europe by the Channel.

During the last 200 years, engineers have put forward numerous proposals for a fixed Channel Crossing, but until recently these all came to nothing. Now at last the Channel Tunnel is a reality. This is the story of one of the most important and exciting engineering feats of the twentieth century.

1. Take the M20 motorway and look for the signs to the Channel Tunnel. Leave the motorway at Junction 11A for the terminal. Eurotunnel's cross-Channel service is called **Le Shuttle**.

2. If you haven't already bought a ticket, there's no problem. You can pay at the toll booth by cash, credit card or cheque.

3. Trucks are separated from other vehicles before the tolls, and travel on separate shuttles.

4. After the toll booth, you can stop for a snack and visit the shops in the Passenger Terminal, or else head for a shuttle.

5. Travelling under the Channel means you are crossing an international frontier — so you'll need your passport.

6. All passport and customs checks for Britain and France are carried out at the departure point, saving time at the other end.

7. After passing through customs, head for the allocation area. Drive down the loading ramp and on to the boarding platform.

8. From the boarding platform, an attendant directs you on to the shuttle. You then drive through the carriages until you are told to stop.

9. Turn off the engine and put the handbrake on. Within eight minutes or so, you're away. The shuttle is well-lit, spacious and air-conditioned. There are toilets in every third carriage.

10. During your journey, you get up-to-the minute information from visual displays and from Radio Le Shuttle. Then, just 35 minutes later, the shuttle comes to a halt.

11. On arrival, attendants direct you to the front of the shuttle and out on to the exit ramp. From here you carry straight on to the motorway. Remember to drive on the other side of the road!

THE TUNNEL IN ACTION

The Channel Tunnel carries four main types of traffic. Cars and coaches travel in passenger-vehicle shuttles, and heavy goods vehicles (HGVs) in freight shuttles. These services are operated by Eurotunnel under the brand name Le Shuttle. In addition, passenger and freight through-trains, operated by the national railways and other railway companies, also travel through the Tunnel.

In fact, the Channel Tunnel is not one tunnel but three. There are two rail tunnels. Traffic from Britain to France travels through the northern tunnel, while the southern tunnel carries traffic from France to Britain. A third, smaller **service tunnel** allows engineers to travel along the system without stopping the traffic.

At peak times, trains thunder along the Tunnel every three minutes at up to 160km/h (100 mph). There can be seven or more trains in each rail tunnel at any given time. Thousands of vehicles and tens of thousands of people pass through the Tunnel every day. The Tunnel is in operation 24 hours a day, seven days a week, 52 weeks a year in virtually all weathers.

THE CHANNEL TUNNEL

◆ **Tunnel length:** 50.45km (31.35 miles) overall. 38km (24 miles) under the sea

◆ **Journey time:** 35 minutes from terminal to terminal, of which 26 minutes is spent travelling through the Tunnel at peak times

◆ **Shuttle frequency:** up to 4 departures per hour at peak times

◆ **Depth beneath seabed:** Average: 45m (148ft) Maximum: 75m (246ft)

Northern rail tunnel

Service Tunnel Transportation System (STTS) vehicle

Passenger-vehicle shuttle

le Shuttle

CONNECTING TUNNELS

The three tunnels are connected together by cross-passages every 375 m (1230ft). These give engineers access from the service tunnel into the rail tunnels to carry out maintenance work. The two rail tunnels are also joined every 250m (820ft) by **piston relief ducts**. Open valves in these ducts allow the air pushed down the rail tunnel in front of the speeding trains to discharge harmlessly into the other rail tunnel. The valves are only closed during maintenance work. Air pressure in the service tunnel is kept higher than in the rail tunnels. This means that the atmosphere in the service tunnel can be kept clear in the unlikely event of smoke being present in the rail tunnels.

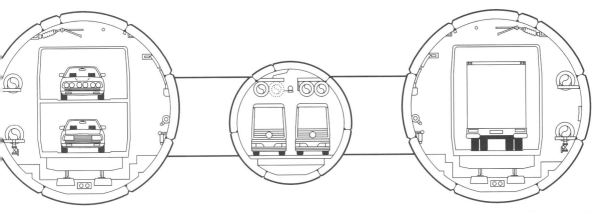

A cross-section view of the three tunnels. A double-deck passenger shuttle is shown in the rail tunnel on the left.

Two Service Tunnel Transportation Vehicles (see page 20) pass each other in the service tunnel.

A freight shuttle is travelling through the right-hand rail tunnel. Each of the wagons can carry one truck of up to 44 tonnes.

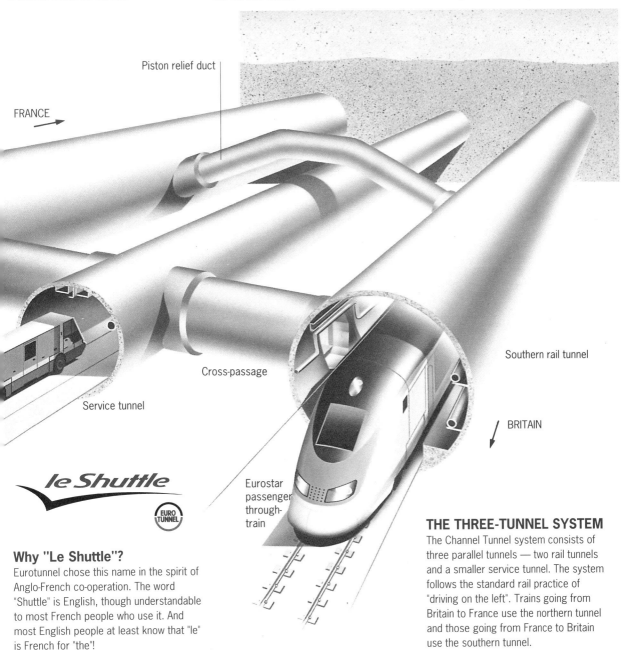

Piston relief duct

FRANCE

Cross-passage

Service tunnel

Southern rail tunnel

BRITAIN

le Shuttle

EURO TUNNEL

Eurostar passenger through-train

Why "Le Shuttle"?

Eurotunnel chose this name in the spirit of Anglo-French co-operation. The word "Shuttle" is English, though understandable to most French people who use it. And most English people at least know that "le" is French for "the"!

THE THREE-TUNNEL SYSTEM

The Channel Tunnel system consists of three parallel tunnels — two rail tunnels and a smaller service tunnel. The system follows the standard rail practice of "driving on the left". Trains going from Britain to France use the northern tunnel and those going from France to Britain use the southern tunnel.

TAKING ON THE TRAFFIC

For cars, coaches and trucks, the shuttle service through the Tunnel is like a moving motorway. They drive on at one end and off at the other.

The rail tracks that run through the Tunnel are linked to the main British and French rail networks. Passenger and freight trains are therefore able to go straight through the Tunnel without stopping at the terminals. The high-speed Eurostar trains connecting London, Paris and Brussels are scheduled to start in the second half of 1994.

Many of the freight trains will be container trains travelling between UK and Continental industrial centres. Later, there will be night sleeper services, as well as trains to the Netherlands, Germany and other parts of Britain.

The Eurostar locomotives are modified versions of the **TGV** (*train à grande vitesse*, or high-speed train) already in use on the French rail system. This can cruise at speeds of up to 300 km/h (185mph) on special tracks. In May 1990, one achieved a world record speed of 515km/h (322mph).

Above: The Eurostar Train is a version of the TGV that has been specially adapted to run through the Tunnel. In Britain, however, the existing tracks to London will not allow it to run at speeds of over 160km/h (100mph).

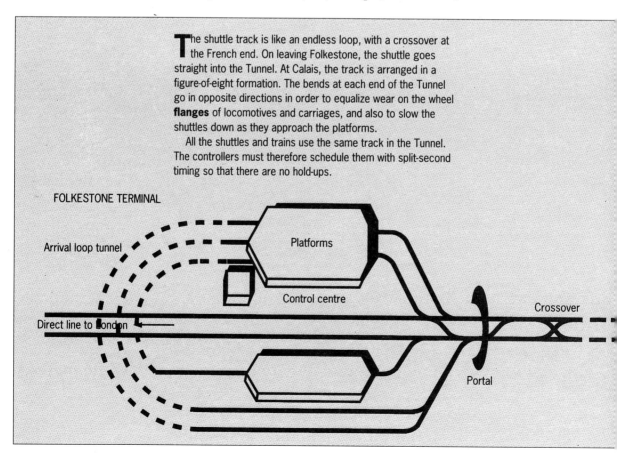

The shuttle track is like an endless loop, with a crossover at the French end. On leaving Folkestone, the shuttle goes straight into the Tunnel. At Calais, the track is arranged in a figure-of-eight formation. The bends at each end of the Tunnel go in opposite directions in order to equalize wear on the wheel **flanges** of locomotives and carriages, and also to slow the shuttles down as they approach the platforms.

All the shuttles and trains use the same track in the Tunnel. The controllers must therefore schedule them with split-second timing so that there are no hold-ups.

FOLKESTONE TERMINAL

Arrival loop tunnel

Platforms

Control centre

Direct line to London

Crossover

Portal

Above: Le Shuttle locomotives have been specially built for use in the Tunnel. Each shuttle has two locomotives — one at each end to provide maximum power. Each pair of locomotives will travel through the Tunnel 20 times a day, with a shuttle averaging 760m (almost half a mile) long.

Right: HGVs are loaded on to the freight shuttles via loading wagons at the end of each section. The trucks board the shuttles in two streams of up to 14 vehicles each to speed up the loading process.

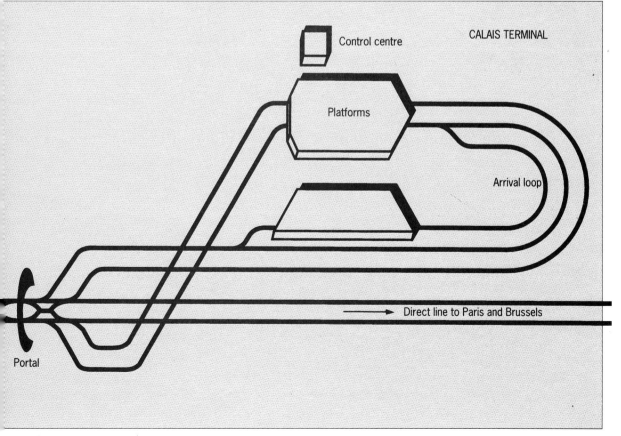

Control centre

CALAIS TERMINAL

Platforms

Arrival loop

Direct line to Paris and Brussels

Portal

THE SHUTTLE LOCOMOTIVES

Shuttle locomotives — which always operate in pairs — are exceptionally powerful engines. They have to be capable of pulling a 2400-tonne train up a gradient of 1 in 90 (that is, a slope that rises one metre in every 90 metres of distance travelled). This is steep in railway terms, and the locomotives have to be able to do it at speeds of up to 140km/h (87mph).

The specially-designed electric locomotive picks up its power from a 25,000 volt (25Kv) overhead power cable (the **catenary**). Its six motors develop a maximum power of 5.76 megawatts, or 7600 horsepower. That's over 100 times the power of an average family car. All six axles, arranged on three two-axle **bogies**, are separately powered. However, the power has to be limited up to speeds of 65km/h (40mph), otherwise the wheels will slip.

BRAKING SYSTEMS

The shuttle locomotive has two main braking systems. One is a standard friction-type system operating directly on the wheels, like that in a car.

The other is called **electrical regeneration**. In this system, instead of the shuttle's electric motors powering its wheels, the wheels power the motors, so generating electricity instead of using it. The power thus generated is returned to the catenary. The shuttle carriages are also fitted with disc brakes.

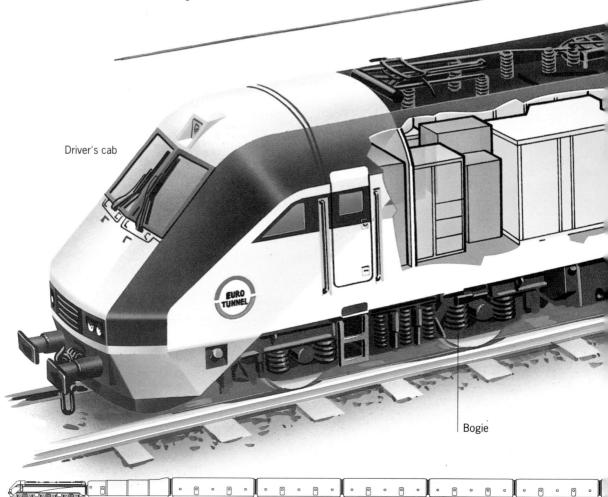

Driver's cab

Bogie

THE SHUTTLE LOCOMOTIVES

◆ **Maximum speed:**
160km/h (100mph)

◆ **Normal operating speed:**
140 km/h (87mph)

◆ **Power:**
5.76MW (7600 horsepower)

◆ **Weight:**
132 tonnes

◆ **Track gauge:**
1.435m (56.5 in)

◆ **Wheel diameter:**
1.250m (49in)

The **pantograph**, which picks up the electric power from the overhead cable, or catenary, has small wings on it to give it lift. This helps push the pantograph against the cable as the locomotive speeds along the track.

DIESEL-ELECTRIC LOCOMOTIVES

Eurotunnel also operates five diesel-electric locomotives which are normally used for maintenance work. They can also be used to pull the shuttle in the unlikely event of a total power failure in the tunnels.

Catenary

Traction converter

THE LOCOMOTIVE'S BODYSHELL

The shuttle locomotive's 30-tonne bodyshell is made up of over 5000 separate parts. The smallest are tiny ribs just 25mm (1in) across. The largest is the single sheet of steel, 16m (52ft) long, which makes up the side of the locomotive.

To ensure that the side panels were smooth and flat, they were laid on a jig and pulled lengthways with a force of 40 tonnes. This stretched the steel by between 6mm and 8mm (0.25 - 0.3in).

The steel for the locomotive varies in thickness between 2mm (0.08in) and 60mm (2.5in). All 5000 pieces were assembled into a single shell, and held together with over 10,000 welds. The shell was put on a special test rig and subjected to punishing stresses and strains.

ROLLING STOCK

The two types of shuttle — passenger and freight — are entirely separate. Passenger vehicle shuttles carry cars, coaches and motorcycles. Freight shuttles carry trucks. Le Shuttle's **rolling stock** is designed to have a working life of 30 years.

Passenger-vehicle shuttles

Travellers on passenger-vehicle shuttles stay with their vehicles during the journey. The air-conditioned passenger carriages are completely enclosed, with fire doors at each end. There are, however, access doors through these barriers so that passengers can walk along the shuttle.

Cars and other vehicles under 1.85m (6ft) high travel in double-deck carriages. These carriages can carry ten cars — five on each deck. Some double-deck carriages have stairs between the upper and lower decks, and there are toilets in every third carriage.

Motorcyclists also travel in double-deck shuttles. They park their bikes in a special section of the loading carriage, and travel in a separate compartment. Cyclists travel in a special bus, with a cycle trailer attached.

Coaches and other high-sided vehicles travel in single-deck carriages. These can either carry one coach or two cars with caravans or trailers.

There are no fewer than 308 electric motors on each passenger shuttle's air-conditioning system. The air flow is so rapid that the air inside the carriage can be completely changed in 70 seconds. The air-conditioning system — along with the ducts, lighting and fire- and fume-warning systems — are constantly monitored by the **train captain**.

The carriages are grouped in sections, called **rakes**. Each passenger vehicle shuttle consists of one rake of twelve single-deck carriages, and another of twelve double-deck carriages, both with loading/unloading carriages at each end. The single-deck rake is always at the front of the shuttle.

Vehicles drive on to Le Shuttle through loading wagons. Double-deck loading wagons have an internal ramp so that car can be driven up to the top deck.

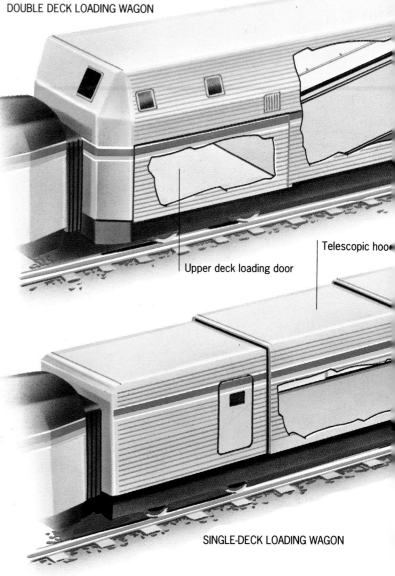

DOUBLE DECK LOADING WAGON

Telescopic hoo

Upper deck loading door

SINGLE-DECK LOADING WAGON

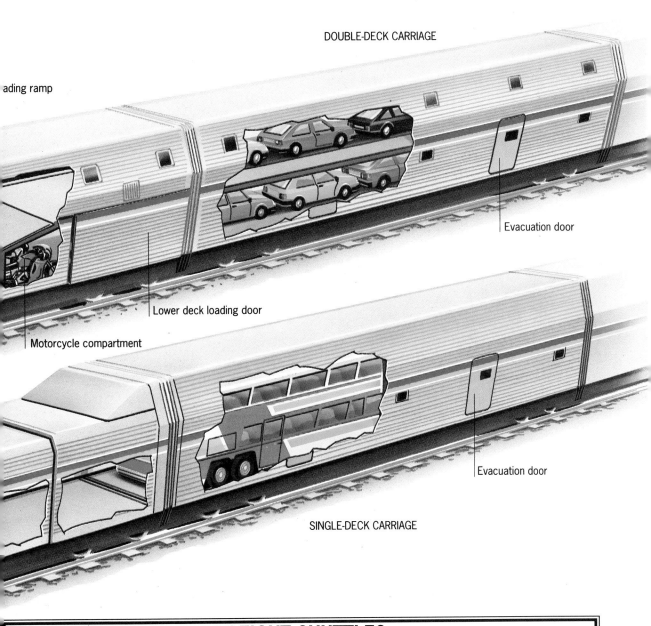

DOUBLE-DECK CARRIAGE

ading ramp

Evacuation door

Lower deck loading door

Motorcycle compartment

Evacuation door

SINGLE-DECK CARRIAGE

FREIGHT SHUTTLES

Freight shuttles have 28 carrier wagons, arranged in two sections of 14 wagons each. Both sections have a loading wagon at one end and an unloading wagon at the other. The carrier wagons, which can each carry one truck of up to 44 tonnes, have a semi-open framework to reduce weight.

The HGVs drive from the platform and on to one of the flat loading wagons.

These have side flaps which are lowered on to the platform to make a ramp. The HGVs are driven forward through the shuttle to the first available carrier wagon. Once the vehicle is in position on the carrier wagon, the driver applies the brakes and switches off the engine. Drivers of refrigerated trucks can connect up their vehicles' cooling systems to power points in the freight wagons to

keep their freight refrigerated during the journey. Drivers of HGVs do not stay with their vehicles. Instead, they are taken by minibus to the air-conditioned Club Car at the front of the shuttle, immediately behind the locomotive. Here they can relax and enjoy a meal. At the end of their journey, they rejoin their trucks, and leave the shuttle via the unloading wagon at the front of each section.

THE DRIVER'S CAB

Imagine that you are driving a passenger-vehicle shuttle. Each shuttle is powered by two locomotives, one at each end. As the driver, you sit in the left-hand seat of the cab in the front locomotive. The train captain, who is responsible for the safety of the shuttle and its passengers, sits in the right-hand seat of the rear locomotive and communicates with you by intercom. The shuttle has a crew of six plus the driver and captain.

The panel in front of you has all the controls needed to drive the shuttle. When the train captain tells you to go, you release the brake, and then you turn the rotating Selector Controller to "forward". Now engage the Power Controller lever next to it — and the shuttle begins to move. Soon, the shuttle is gathering speed. In just a few minutes it reaches the portal and goes into the Tunnel itself. A specially-designed signalling system called TVM430 controls your speed. This French system provides information via indicators in the cab, instead of trackside signals. (TVM stands for *Transmission Voie Machine* — track to train transmission.)

The speed at which you should travel flashes up on lights in front of you. Top speed is 140 km/hr (85mph). If it is safe to drive at this speed, the figure 140 is displayed in black on green. If you have to slow down, the display first starts to flash as a warning, and then the new lower speed is displayed. If you go too fast, the shuttle's brakes go on automatically.

The **VDU** in front of you automatically warns you if any of the shuttle's equipment is not working properly. The train captain uses two VDUs to monitor what is happening in every part of the shuttle. The controls in the cab are laid out for ease of use whether you are standing up or sitting down.

"SEGMENT FLICKER"

High-speed trains are already designed so that the driver cannot see any track closer than 20m (65ft) ahead. This is because the flickering effect of the sleepers disappearing rapidly under the train could in theory hypnotise the driver. In the case of Le Shuttle, the pattern of tunnel sections flashing by could have the same effect. Le Shuttle's cabin is therefore designed with small windows so that the driver cannot see them close by.

CONTROLLING THE TRAFFIC

There are different systems for controlling road and rail traffic. The road traffic controllers work from the top floor of the control centre at each terminal. They have a direct view on to the terminal, and get additional information from video cameras. After leaving the frontier controls, vehicles are automatically counted and allocated to different lanes before boarding their shuttles.

Rail traffic through the Tunnel is normally controlled from the Folkestone control centre. In the rail control room, the controllers sit at three rows of desks, with a **mimic board** showing the entire system on a curved wall in front of them. This indicates the exact position of every train and shuttle in the tunnels and on the terminals. Signals are sent direct to the driver's cab, rather than through a conventional trackside system. If the driver ignores a signal, the train will automatically slow down or stop.

If there is a fault at the Folkestone centre, the Calais rail control room takes over immediately.

Below left: The control centre building at the Folkestone Terminal. Under normal circumstances, the entire Channel Tunnel rail system is controlled from Folkestone.

Right: The "mimic board" in the rail control centre at Folkestone. This shows the position of all trains and shuttles that are in the tunnels at any one time. The control and communications systems are designed to continue operating in the event of a power cut.

Below right: The view from the control centre building at the Calais Terminal, which acts as a back-up in case of emergency, and is fully staffed at all times. From time to time overall control of the system is switched from Folkestone to the Calais control centre — just to check that everything is working properly.

MAINTENANCE

The Channel Tunnel is a complete, self-contained rail system, with nine passenger-vehicle shuttles, eight freight shuttles, and 38 shuttle locomotives, making a grand total of no fewer than 567 vehicles.

There are two maintenance workshops. The main one, at Calais, occupies 10,000 sq m (100,000 sq ft). There are four tracks, with inspection pits under the rails, and maintenance platforms at floor levels. There is a smaller workshop — just 750 sq m (8000 sq ft) — at Folkestone.

As far as possible the aim is to anticipate problems and deal with them before they occur. To this end, each locomotive and its rake, or half-shuttle, are inspected and serviced at weekly intervals. The weekly maintenance check takes about eight hours for a passenger-vehicle shuttle and about six hours for a freight shuttle, which is not as complex.

Most of the work takes place at night when there is less traffic. The weekly programme is supplemented by annual maintenance checks. In addition, a team of "trouble-shooters" is based at each terminal. They can undertake minor repairs to a shuttle while it is being loaded or unloaded.

Above: The huge maintenance sheds at Calais operate around the clock to keep Le Shuttle working.

Right: A shuttle locomotive awaiting inspection at the Calais maintenance workshop. Eurotunnel employs around 260 maintenance engineers, almost all of them in Calais.

Left: Routine inspections of the underside of the shuttles enable engineers to spot potential problems before they happen.

SERVICE TUNNEL VEHICLES

Shuttles and trains are not the only vehicles to use the Tunnel. There are others which as a passenger you are unlikely to see in operation, as they are used in the service tunnel. Their official name is Service Tunnel Transportation System vehicles, but they are usually known as **STTS** vehicles for short. They run on rubber tyres, and are powered by diesel engines. Their exhaust gases are specially treated to prevent fumes building up in the tunnels.

The STTSs are designed for use both during routine maintenance and also in emergencies. Under normal circumstances they are steered by an electronic guidance system that follows wires buried under the floor of the service tunnel. Two vehicles can then automatically pass each other.

Below: The STTS vehicles have been designed with a cab and engine at each end. This feature is necessary because they cannot turn around in the service tunnel. The tunnel is just wide enough for two to pass each other.

Right: The STTS vehicles serve several purposes. While the driving unit stays the same, different modules, or "pods", can be slid in and out of the body.

The three different modules are: (1) maintenance, (2) fire-fighting and (3) ambulance. Pods can be switched in just a few minutes.

SERVICE TUNNEL TRANSPORTATION SYSTEM (STTS) VEHICLES

◆ **Length:** 10.2m (33.5ft)

◆ **Width:** 1.5m (5ft)

◆ **Cruising speed:** 50km/h (30mph)

◆ **Diameter of service tunnel:** 4.8m (16ft)

EARLY ATTEMPTS

The Channel is only 34km (21 miles) wide at its narrowest point between Dover and Calais. However, until comparatively recently, crossing it was a slow and uncomfortable business. Engineers have long dreamed of building a fixed crossing, independent of wind and tide.

The first serious proposal was put forward in 1802, in a brief period of peace during the Napoleonic wars. It was the brainchild of a French engineer named Albert Mathieu-Favier. His plan envisaged that passengers would travel in horse-drawn coaches in an undersea tunnel ventilated by huge iron chimneys. But resumption of the war between Britain and France put paid to Mathieu-Favier's scheme.

Later in the nineteenth century, another Frenchman, M.J.A. Thomé de Gamond, proposed various schemes for a rail tunnel. In 1867, he presented a further scheme at the Great Exhibition in Paris. His tunnel would have run via an artificial island on the Varne Bank in mid-Channel.

A Channel Tunnel Company was formed in 1876, and the following year work began at Sangatte, near Calais. The tunnel boring machine was developed from designs made by two military engineers called Frederick Beaumont and Thomas English. By 1882, a tunnel going 1.85km (1.15 miles) out to sea had been bored from Shakespeare Cliff when fear of invasion again put a stop to the work.

The 1880s scheme was resurrected in the 1920s, but did not progress beyond trial borings. Work was not to restart in earnest until 1974. A year later the project was halted once more — on economic grounds. It was not until the 1980s that the dreams of the pioneers began to become reality.

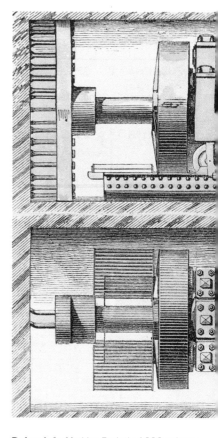

Below left: Mathieu-Favier's 1802 scheme for a coach service beneath the Channel incorporated huge ventilation chimneys.

Below: A trial tunnel bored at the start of the 1880s scheme. Despite never having been lined, the tunnel is still intact and dry over 100 years later.

THE BEAUMONT-ENGLISH TUNNELLING MACHINE

A contemporary plan (below) of the boring machine used in the 1880s scheme. It was developed from a design by Colonel Frederick Beaumont, which had been improved by Thomas English. It was 9m (30ft) long and powered by compressed air. An improved Beaumont machine was also used on the French side.

Bottom: An 1880s cartoon from the US magazine *Puck*, satirizing British military objections to the building of a tunnel. General Wolseley, one of the scheme's main opponents, rides the retreating lion.

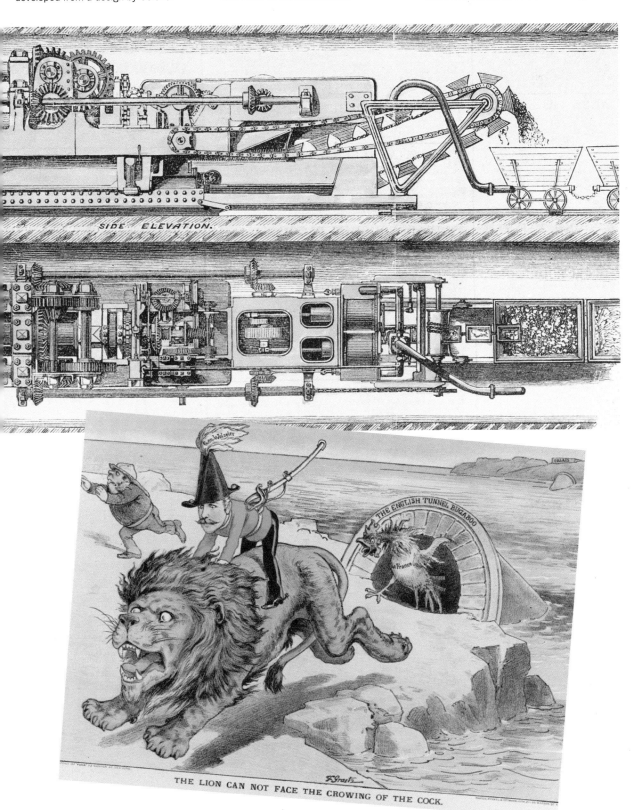

SIDE ELEVATION.

THE LION CAN NOT FACE THE CROWING OF THE COCK.

THE CHALLENGE

Not all of the proposals for a fixed Channel crossing assumed that it would be a tunnel. Even some of the schemes with which the design for Eurotunnel had to compete involved a bridge for some or all of the route.

During the nineteenth century, engineers had come up with all kinds of ideas for a permanent crossing. One suggested laying a huge metal tube along the sea bed through which trains could run. Another recommended the use of submarines to build the stone piers of a great bridge. Yet another engineer wanted to construct a huge land bridge by simply filling in a section of the Channel — although he proposed to build canals to let ships through!

THE WORLD'S LONGEST TUNNELS

The Channel Tunnel is the world's longest undersea tunnel system. It runs for 38km (24 miles) under the sea. The average total length of the three tunnels is 50.45km (31.35 miles) from portal to portal.

The Seikan Tunnel in Japan is longer overall at 53.85km (33.46 miles), but only 23.3km (14.3 miles) of its length is under the sea. It connects Japan's main island — Honshu — with the island of Hokkaido.

Sydney Harbour Bridge (arch)

Forth Rail Bridge (cantilever)

Golden Gate Bridge (suspension)

THE RIVAL SCHEMES

In September 1981, the then British Prime Minister Margaret Thatcher and President Mitterrand of France announced their joint intention that a fixed link should be built between their respective countries. On 12 February 1986, they signed the Treaty of Canterbury, which paved the way for both countries to pass the necessary laws.

During the previous year, the two governments had held a competition for the best plan. By the closing date of 31 October 1985, nine plans had been submitted. Five of these were rejected almost immediately. An Anglo-French Committee was formed to study the remaining four.

One approach was an enormously long suspension bridge across the Channel. That, however, would have represented a hazard to shipping in the busiest sea lane in the world. It might also have had to be closed in severe weather. And certainly no bridge of that size had ever been attempted.

A second proposal involved a bridge combined with a huge tube, carrying a road, that would run along the sea bed across the central section of the Channel. A joint road and rail tunnel was also proposed, but this would have been very expensive to build. Four tunnels would have had to be dug. Then there would be the problem of vehicles breaking down and blocking the tunnels. Finally, any proposal involving a road tunnel would have needed a super-efficient ventilation system to remove vehicle exhaust fumes.

In the end, it was decided that the most practical solution was a rail link, with vehicles being transported though the tunnel on shuttle trains.

Left: The Seikan Tunnel is the longest tunnel in the world. It dips to 240m (787ft) below sea level, and 100m (328ft) below the sea bed. The main tunnel was built between June 1972 and January 1983. The first (rail) traffic ran through it in March 1988.

Seikan Rail tunnel

TYPES OF BRIDGE

Arch bridges are one of the simplest types of bridge. The weight of the arch is carried down the supporting pillars. With its 500m (1650ft) span, the steel-arch Sydney Harbour Bridge is an example of this type.

Cantilever bridges consist of projecting beams joined in the middle, sometimes by another beam called the suspended span. They are supported by piers, and attached to supports on either side of the river or other obstacle. The Forth Rail Bridge in Scotland is a cantilever bridge.

Suspension bridges are used where exceptionally wide spans are needed. In bridges of this type, the roadway hangs from rods attached to cables which are supported by towers. The Golden Gate Bridge in San Francisco, USA, is one of the world's most famous suspension bridges.

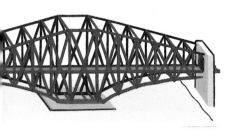

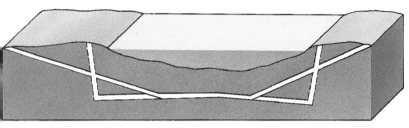

CHOOSING A ROUTE

One reason why engineers have been so enthusiastic about the concept of a tunnel for so long is that the rock beneath the Channel is ideal for tunnelling. At the Straits of Dover — the Channel's narrowest point — there is a layer of **chalk marl** that runs almost all the way across.

Marl is a mixture of chalk and clay. It is soft, so it is easy to tunnel through and unlikely to crack or crumble. Unlike chalk, it is **impervious** — more or less waterproof. It is also strong enough to avoid collapsing when holes are bored through it. When the 1880s tunnel at Shakespeare Cliff was reopened a century later, it was still dry and in good condition. (It has since been resealed for safety reasons.)

Below: A geological cross-section (not to scale) through the floor of the Channel. It shows how the Tunnel follows the layer of chalk marl. In surveys for early schemes, engineers described this rock as having the consistency of very hard cheese — they found that it was just possible to cut it with an ordinary penknife.

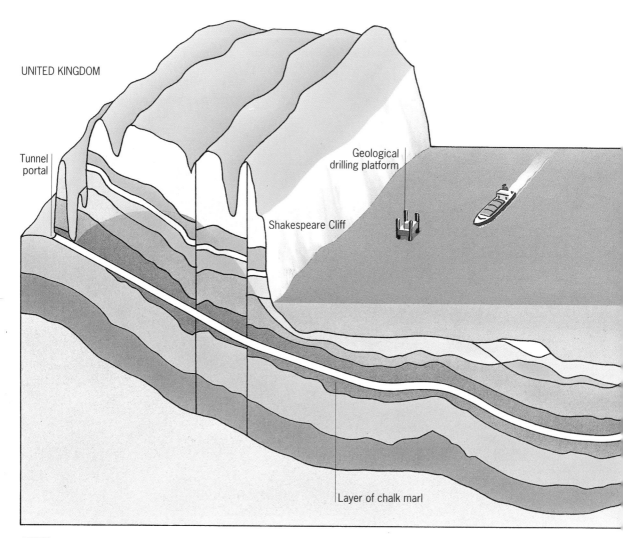

UNITED KINGDOM

Tunnel portal

Geological drilling platform

Shakespeare Cliff

Layer of chalk marl

DRILLING FOR CORE SAMPLES

Ever since the Channel Tunnel was first proposed, geologists have been collecting rock samples from beneath the sea. In the nineteenth century, Aimé Thomé de Gamond even lowered himself from a rowing boat to examine the sea bed. His feet were weighted down with bags of stones, and he carried inflated pigs' bladders around his waist to help him regain the surface. He has left a graphic description of being attacked by conger eels on one occasion.

Nowadays, investigations are carried out in a much more scientific way. Small explosions are set off and the echoes of the seismic waves bouncing off the rocks under the sea floor are measured. Platforms like this one were used to drill core samples from depths up to 80m (260ft) below the sea bed.

CHALK MARL LAYER

◆ **Thickness:** 70 - 75m (230 -250ft)

◆ **Slope (British side):** 5°

◆ **Slope (French side):** 20°

◆ **Age:** 65-135 million years

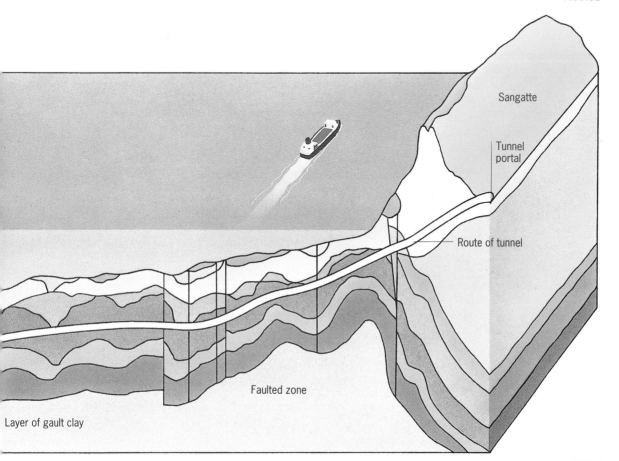

FRANCE

Sangatte

Tunnel portal

Route of tunnel

Faulted zone

Layer of gault clay

THE TUNNEL BORING MACHINES

O nce the route had been chosen, huge tunnel boring machines (**TBMs**) had to be built. The biggest was 8.78m (28.17ft) in diameter and weighed 1575 tonnes. With the **service train** connected up behind it, it was 260m (853ft) in length — longer than two football pitches joined end to end.

The teeth of the TBMs' revolving cutting heads were made of a very hard material called tungsten carbide. They were able to cut their way through the chalk marl at the rate of up to one kilometre (0.62 miles) per month.

The cutting head was pushed forward into the chalk by powerful hydraulic rams. The rams in turn pushed against the follow-on gripper section. This was locked in place by pads pressing against the tunnel walls. As the cutting head moved forward, the telescopic skin that joined it to the gripper section extended. After each tunnel lining ring was installed, the gripper section was driven forward to catch up with the cutting head by rams pressing against the ring.

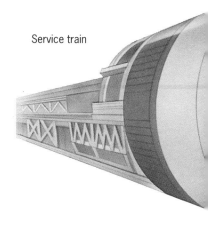

Service train

Left: The British TBMs were brought to Dover by sea in sections. The sections were taken to Shakespeare Cliff by truck and lowered down the **access shaft**. They were finally assembled underground. Four of the British TBMs were made in Scotland, and two in Chesterfield, England. Here, sections of the south landward rail tunnel TBM are delivered to Dover docks in August 1989.

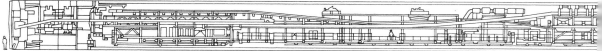

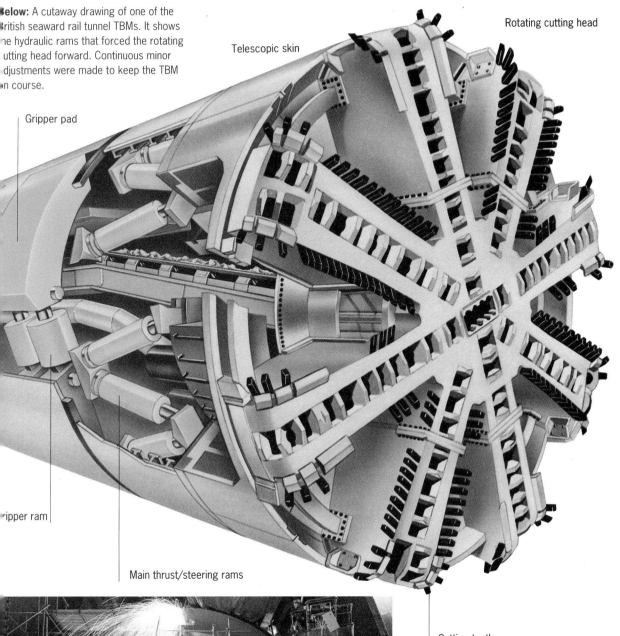

Below: A cutaway drawing of one of the British seaward rail tunnel TBMs. It shows the hydraulic rams that forced the rotating cutting head forward. Continuous minor adjustments were made to keep the TBM on course.

Telescopic skin

Rotating cutting head

Gripper pad

Gripper ram

Main thrust/steering rams

Cutting teeth

Left: The British TBMs were taken to a specially-excavated chamber near the foot of the access shaft for assembly. Here, sections of the north seaward rail tunnel TBM are being welded together.

The French TBMs were partly assembled on the surface before being lowered down the wider access shaft at Sangatte.

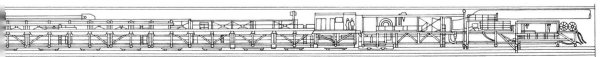

WORK STARTS

Before the TBMs could start work, access shafts had to be dug. These led from the surface to the level of the Tunnel. In Britain, a small access shaft, 10m (33ft) in diameter, was dug from the top of Shakespeare Cliff to the Tunnel workings. At Sangatte, where it was first necessary to drain a large area around it, a much bigger access shaft was built, 55m (180ft) in diameter and 75m (246ft) deep. The TBMs were lowered down the access shafts in sections. Assembly was completed underground.

There were eleven TBMs in all — for twelve tunnels. Three TBMs — one for each tunnel — set out under the sea from each side of the Channel. Another three went inland from Shakespeare Cliff, towards the Folkestone Terminal. In France, only two TBMs were needed for the inland tunnels from Sangatte towards the Calais Terminal. One bored the service tunnel, the other both rail tunnels (see pages 36-37).

Above: A 630-tonne crane lowers part of the cutting head of one of the French TBMs down the huge access shaft at Sangatte. The access shaft was roofed so that work could continue in all weathers.

Left: The cutting head of a British rail tunnel TBM is lowered down the access shaft at Shakespeare Cliff. The shaft was just big enough to let the machinery through.

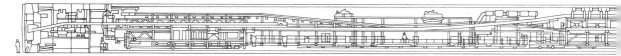

Below top: The interior of a section of one of the French seaward rail tunnel TBMs.

Below centre: The cutting head of one of the French machines is moved along a seaward rail tunnel before final assembly.

Below right: A French TBM is assembled at the cutting face of one of the seaward running rail tunnels.

EUROTUNNEL AND TML

In 1986, the British and French governments chose the scheme put forward by the Channel Tunnel Group/France-Manche (CTG/FM) for the creation of a fixed link between the two countries. At this point it was realised that it was necessary to separate the roles of the contractor who actually built the Tunnel and the operator who owned and ran it. The ten construction companies (five French, five British) that comprised CTG/FM formed themselves into a bi-national engineering and construction contractor, called Transmanche-Link (TML). A contract was negotiated between TML and the newly formed holding company, Eurotunnel (a partnership comprising Eurotunnel plc in Britain and Eurotunnel SA in France). Once the Tunnel was finished, it was handed over to Eurotunnel as the owner/operator.

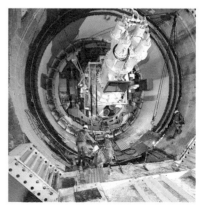

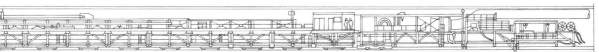

THE TUNNEL LINING

Behind the TBM, the tunnel walls were lined with rings made up of concrete segments. These were brought up through the **supply train** before being installed. Each ring was then locked into place with a smaller, wedge-shaped "key" segment. The tiny 20mm ($\frac{3}{4}$ in) gap between the ring and the rock face was then sealed with cement **grout**. It took about 20 minutes to fix each segment in place.

In both Britain and France, special factories were set up to make the tunnel linings. The French factory was at Sangatte; the British one at the Isle of Grain in Kent, some 100km (60 miles) away — there was not enough room on site. The British tunnel linings each consisted of eight segments plus the "key" segment. The French tunnel linings consisted of six segments plus the key. Cast-iron linings were used instead of concrete in areas of weaker rock and at cross-passage junctions.

Above: Hand finishing some of the 400,000 segments for the French tunnel linings before heat treatment in a curing tunnel. After curing, the segments were left to harden before use.

Left: The tunnel lining segments were reinforced with a framework of steel rods. Here, reinforcement "skeletons" are being made in the French factory at Sangatte.

Below: Lowering cured segments down the vast access shaft at Sangatte using a 60-tonne crane.

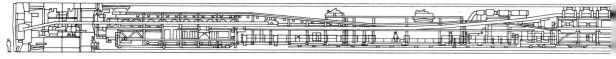

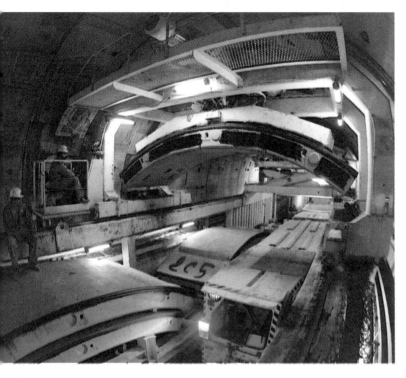

Above: Concrete lining segments are fitted in one of the UK seaward rail tunnels.

Left: Segments being lifted from the supply train to the TBM in a French tunnel.

Below: A "key" segment is pressed into place in a British landward rail tunnel.

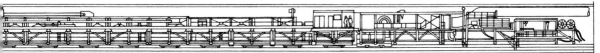

KEEPING ON COURSE

Keeping the giant TBMs on course was vital. Even the slightest deviation — a fraction of one degree — would have meant that the British and French tunnels did not meet up under the middle of the Channel. The problem is that, once underground, it is not easy to tell exactly where you are. Even the latest satellite guidance systems are no use.

The only way to keep on course is to survey down the tunnel to work out where you are. The tunnel engineers did so with incredible accuracy, using laser beams. The on-board computer in the cabin reacted to the laser beam and plotted the TBM's exact position. The operator could then adjust the hydraulic rams behind the cutting head to drive the TBM forward on exactly the right course.

Below: The beam of red light on the extreme right of the picture comes from a laser. Computerized laser guidance systems were used to make sure that the TBMs stayed exactly on course.

The cutting head of the TBM rotated at a speed of 2-3 revolutions per minute. Its direction was controlled by eight individually adjustable hydraulic rams. Each of these had a thrust of up to 210 kilograms per square centimetre (3000 pounds per square inch).

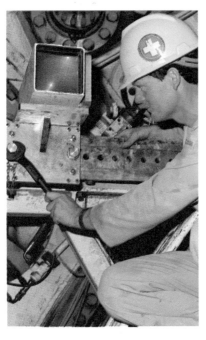

1. A laser beam is aimed along the tunnel so that it shines through control points fixed to the tunnel wall.

2. The beam hits a target at the rear of the cutting head. The exact position is then relayed to the TBM's computer.

3. From the TBM's control cabin, the operator adjusts the rams to keep the machine exactly on course.

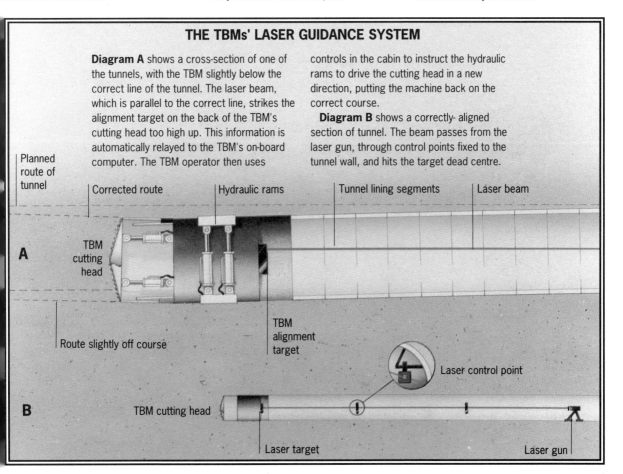

THE TBMs' LASER GUIDANCE SYSTEM

Diagram A shows a cross-section of one of the tunnels, with the TBM slightly below the correct line of the tunnel. The laser beam, which is parallel to the correct line, strikes the alignment target on the back of the TBM's cutting head too high up. This information is automatically relayed to the TBM's on-board computer. The TBM operator then uses controls in the cabin to instruct the hydraulic rams to drive the cutting head in a new direction, putting the machine back on the correct course.

Diagram B shows a correctly- aligned section of tunnel. The beam passes from the laser gun, through control points fixed to the tunnel wall, and hits the target dead centre.

Planned route of tunnel

Corrected route

Hydraulic rams

Tunnel lining segments

Laser beam

A

TBM cutting head

Route slightly off course

TBM alignment target

B

TBM cutting head

Laser control point

Laser target

Laser gun

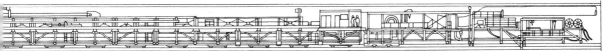

BEHIND THE MACHINE

The front section of the **service train** carried the pumps that supplied hydraulic power for the rams that drove the cutting head and erector section. Behind that were the electricity supply units for the pumps and the lights. Ventilator ducts ran through the service train, which also carried air pumps and filter units to trap the dust.

Supply trains, running on temporary tracks laid along the tunnels, brought up the lining segments. These were then lifted into position by overhead gantry cranes. Conveyor belts carried the **spoil** (fragments of rock) from the cutting head, through the service train, to the supply train. This took the spoil out of the tunnel.

Below: When the first of the French landward rail tunnels was completed, the TBM and its service train were turned around to bore the other tunnel. It was the only TBM that finished up where it started — at Sangatte. This photograph shows the support train of the TBM after the machine was turned round.

The service trains were each up to 260m (850 ft) in length. Conveyors on the service train carried away the spoil, and machines hoisted the tunnel lining sections into place. In addition to supplying food and water, the service train even had a mess room in which the crew could eat.

THE COST

At the time of opening, the Channel Tunnel had cost almost £9 billion. About two-thirds of this sum was spent on designing and constructing the system and its rolling stock. The remainder was made up of financing and corporate costs.

Left: The piston relief ducts were excavated with hand tools. Small **roadheaders** were used to dig the cross-passages. On the British side, these were lined with cast-iron rings brought along the tunnel by the supply train. On the French side, concrete linings were used.

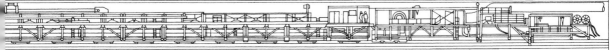

DISPOSING OF THE SPOIL

When the TBMs were running at full speed, spoil was coming out of the British side of the Tunnel at up to 2400 tonnes an hour. Nearly eight million cubic metres had to be disposed of. The total spoil removed from both ends of the Tunnel could have filled Wembley stadium 13 times over. The spoil was carried from the service train conveyors down the Tunnel in **muck wagons**.

On the French side, these wagons were tipped upside down, and the spoil dropped into a huge sump at the bottom of the access shaft at Sangatte. There, the spoil was crushed and mixed with water to form **slurry**. On the top of a hill one kilometre (0.6 miles) away, a dam 730m (2400ft) in length was built. The slurry was pumped up the hill and into the artificial lake formed by the dam. When it had dried out, it was landscaped and grassed over.

On the British side, a sea wall was extended at the foot of the construction site at Shakespeare Cliff. The spoil was then dumped into the resulting artificial lagoons.

Above: The spoil conveyor ran though the middle of the service train.
Left: The cutting face of one of the British landward rail tunnel TBMs.
Below: The spoil conveyor emerging from tunnel workings below Shakespeare Cliff.

Below left: The lower site at Shakespeare Cliff. The sea wall is under construction. At the far end of the site are the artificial lagoons into which the spoil was dumped.

Below centre: At Sangatte the spoil wagons were rotated and the spoil tipped into a sump. Water was then added to the spoil to form slurry.

Below right: The slurry was pumped from Sangatte (top left of picture) to a lagoon at Fond Pignon. There, it was left to dry out before being grassed over.

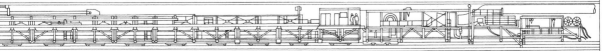

KEEPING UP SUPPLIES

During the construction period, there were six tunnels on the British side alone — three landward and three seaward. Work was going on in five of these at the same time. (Work on one of the landward rail tunnels was not started until the landward service tunnel was completed.)

During the busiest period, trains on the British construction railway travelled a distance equivalent to four times around the world. There were about 1000 items of rolling stock. Of these, two-thirds were **flatbeds** transporting supplies, and one-third were muck wagons which each carried about 14 tonnes of spoil.

Right: The base of the French access shaft. Construction staff were taken to their underground places of work in specially-designed wagons called **manriders** (see page 48).

Below: The construction railway control centre at the top of the access shaft at the Sangatte site. Controllers had to keep up with a system that was being extended every day as work on the tunnels progressed. Computerized systems ensured that equipment and material were in the right place at the right time.

Above: The massive stockpile of tunnel lining segments at Shakespeare Cliff. Nearly half a million segments were used to line the British tunnels alone. As there was insufficient room to build a factory locally, the segments were manufactured at the Isle of Grain in North Kent. They were then transported to the Shakespeare Cliff site by rail.

Left: A spoil train behind one of the British rail tunnel TBMs. A narrow gauge was used for the construction railway so that two of the temporary tracks could be fitted into each tunnel.

Above: A supply train in one of the British seaward rail tunnels. This one is loaded with brackets to secure cooling pipes and other equipment to the tunnel walls.

THE CROSSOVER CAVERNS

Deep under the Channel, there are two huge **crossover caverns,** where trains can change from one rail tunnel to the other. When one of the tunnels is closed for maintenance or repairs, the other has to be used as a single-track railway. If one entire tunnel had to be closed, the trains and shuttles would be considerably delayed. However, because of the crossover caverns, only a section of the tunnel need be out of service at any time. During single-track working, the signalling and train control systems ensure that there are no trains coming the other way. To change from one tunnel to the other, trains have to slow down to less than 60km/h (38mph).

The crossover cavern are located at about one-third and two-thirds of the total undersea distance between the portals. Their exact positions were determined by the need to have as thick a layer of chalk marl as possible between the caverns and the sea bed. As they divide each rail tunnel into three sections, only one-sixth of the system needs to be closed off during maintenance periods.

The crossover caverns are sometimes described as "cathedrals" because of their size. They are high enough to contain three double-decker London buses stacked on top of each other. Huge sliding doors separate the two tunnels when the crossover is not in use.

The crossover doors
The doors between the two tracks are needed to separate the ventilation system of each tunnel as well as for safety. The doors have a framework of super-strong carbon manganese steel with 5mm (0.2in) of steel cladding, plus 15mm (0.6in) of fire resistant material to prevent fire spreading from one tunnel to the other.

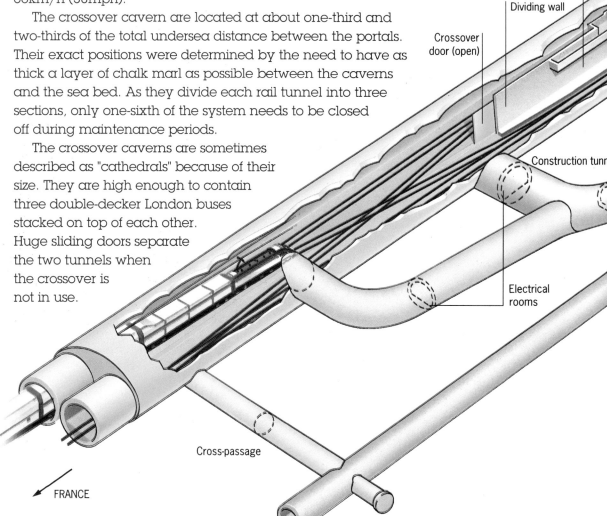

Control and signalling room

Door operation room

Dividing wall

Crossover door (open)

Construction tunnel

Electrical rooms

Cross-passage

FRANCE

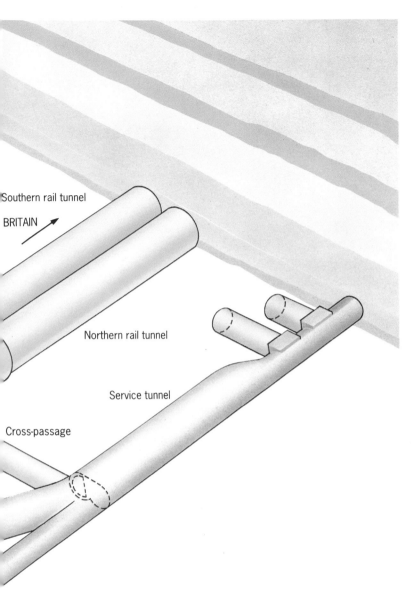

Southern rail tunnel

BRITAIN →

Northern rail tunnel

Service tunnel

Cross-passage

BRITISH CROSSOVER

◆ **Position:** 17km (11miles) from British portal; 7.5km (4.7 miles) from coast

◆ **Depth below sea level:** 77.6m (255ft)

◆ **Depth below sea bed:** 46.6m (153ft)

◆ **Dimensions of cavern:** Length: 156m (512ft). Width: 18m (59ft). Height: 9.5m (31ft)

◆ **Dimensions of doors:** Length (each): 32m (105ft). Height: 6.6m (23ft). Weight (each): 92 tonnes

FRENCH CROSSOVER

◆ **Position:** 15.7km (9 miles) from French portal; 12km (7.5 miles) from coast

◆ **Depth below sea level:** 90m (300ft)

◆ **Depth below sea bed:** 45m (150ft)

◆ **Dimensions of cavern:** Length: 162m (530ft). Width: 18.9m (62ft). Height: 12m (40ft). Dimensions of doors: Length (each): 33m (108ft). Height: 7m (22.5ft). Weight (each): 140 tonnes

Right: The British undersea crossover cavern before the doors were put in place, showing the "scissors" crossover. During construction, tunnels known as "adits" were first dug from the service tunnel. Small tunnels were then excavated along the length of the cavern, and the roof section was dug out. Temporary layers of concrete sprayed on to steel mesh were used to support the excavation, before the permanent lining was added.

On the French side, the chalk was found to be more cracked and variable in quality, which increased the likelihood of water leaking through. Concrete-filled tunnels 2m (6.5ft) in diameter were constructed along the length of the cavern around the previously bored and lined rail tunnels. The remaining chalk could then be safely excavated.

THE TUNNEL'S HIDDEN WORKINGS

A supply of fresh air for the tunnels is provided by huge ventilation plants at Shakespeare Cliff on the British coast and at Sangatte on the French side. Giant fans 2m (6.5ft) in diameter pump 144 cubic metres (5085 cubic feet) of fresh air a second down the service tunnel. That is enough for 20,000 people to breathe.

The passage of the trains heats the air in the tunnels, which then needs to be cooled. Two cooling pipes run through each rail tunnel. Cold water from chilling plants at Sangatte and Shakespeare Cliff circulates through the pipes. The water absorbs heat from the tunnels, and returns to the plants where is is cooled down once more. The British chilling plant has a capacity of over 28MW. The French installation is somewhat smaller, as the French cooling circuits are shorter on the landward side. The two plants keep the temperature in the tunnels between 25° and 35°C (77° to 95°F).

There are also three pumping stations in the Tunnel to pump out any water that seeps in. Two are on the British side, of which one is located at the lowest part of the Tunnel, close to the mid-point. The third pumping station is 8.2km (5.5 miles) from the French coast.

VENTILATION AND COOLING SYSTEMS

◆ **Ventilation fans:**
Weight: 32 tonnes. Capacity 300 cubic metres (10,000 cubic feet) in either direction

◆ **Drainage pumping centre capacity:** 153 litres (34 gallons) per second

◆ **Environmental sensors:**
2.5 million readings are taken each day from 500 environmental sensors in the Tunnel, which measure temperature (for cooling purposes) and carbon monoxide levels.

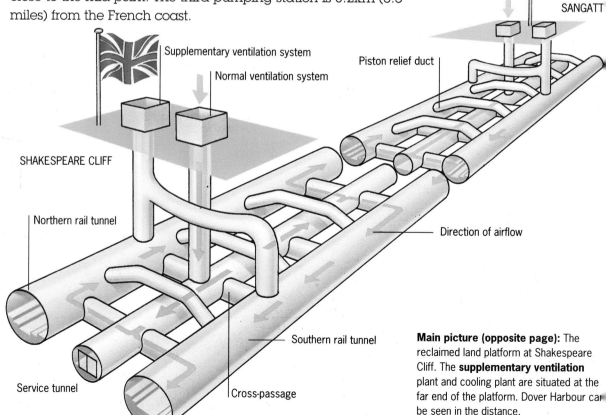

SANGATT

Piston relief duct

Supplementary ventilation system

Normal ventilation system

SHAKESPEARE CLIFF

Northern rail tunnel

Direction of airflow

Southern rail tunnel

Service tunnel

Cross-passage

Main picture (opposite page): The reclaimed land platform at Shakespeare Cliff. The **supplementary ventilation** plant and cooling plant are situated at the far end of the platform. Dover Harbour can be seen in the distance.

Workers take a break in one of the supplementary ventilation fans at Shakespeare Cliff.

Giant ventilation ducts constructed in the shaft at Sangatte lead down to the tunnel.

Inside the cooling plant at Sangatte.

WORKING ON THE TUNNEL

During every eight-hour shift, up to 1000 people were transported to and from their underground places of work by the construction railway. They travelled in three-car wagons called "manriders". Each manrider carried 90 workers. Visitors travelled in a special type of glass-roofed manrider — nicknamed the ***Disneymobile***.

At any one time, there were countless work sites, where cross-passages, piston relief ducts and equipment rooms were dug from the sides of the tunnels. These also needed trains to supply materials and remove the spoil.

Clockwise from opposite page

1. The "Hanging Gallery" at Sangatte, where workers exchanged their everyday clothes for work overalls.

2. Excavating a cross-passage in one of the French rail tunnels.

3. A welder repairs the teeth of a roadheader. This machine was used to expand small tunnels dug by hand to their full diameter.

4. Miners digging one of the piston-relief ducts.

5. Construction workers had to wear safety helmets and self-rescuers at all times. (A self-rescuer includes a face mask and an air filtering system in case of smoke.)

6. Underground workers coming off shift at the top of the access shaft at Shakespeare Cliff.

THE HUMAN ACHIEVEMENT

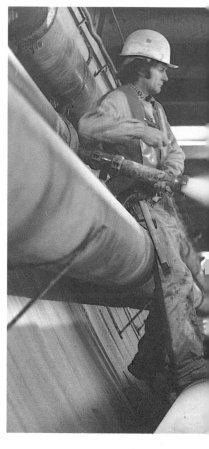

The construction of the Tunnel was a great human achievement. Added to the tunnellers, engineers, surveyors, construction workers and electricians who actually built the system, there were also caterers, clerks, secretaries, managers, and rail and medical staff.

Not surprisingly in a bi-national project, language training was taken very seriously from the start. For safety reasons alone, it was necessary for engineers to have a working knowledge of technical terms in both English and French.

Once the Tunnel was complete, many of the tunnellers who built it went on to work on other major projects such as the new Jubilee underground line in London, the Storebelt in Denmark and the Highland Water project in Lesotho.

Above: A spoil train being hosed down in one of the French tunnels.

Left: TML's senior geologist at work in the British tunnel in May 1989.

Below: A statue of Saint Barbe, the patron saint of miners, stood guard over workers below ground at Sangatte.

Above: The Director of Occupational Health takes atmospheric readings on the French side in May 1990.

Right: Controllers keep a close watch on construction activity from the centre at Shakespeare Cliff.

BREAKTHROUGH

The breakthrough proper took place on 1 December 1990, when the British and French service tunnels were joined up. It was three years to the day after work had begun on the British undersea service tunnel.

A technical breakthrough had occurred a month earlier. Work on the two halves of the service tunnel had been halted when they were about 100m (330ft) apart. A probe 50mm (2in) in diameter was used to confirm the accuracy of the bore. The probe confirmed that the tunnels were correctly aligned to within a few centimetres.

An access tunnel was then dug by hand between the head of the French tunnel and the side of the British tunnel. Following the initial breakthrough, a roadheader was used to excavate the tunnel to its full diameter.

The breakthroughs of the northern and southern rail tunnels took place ahead of schedule on 22 May and 28 June 1991 respectively. The cutting heads of the British TBMs were steered downward, to enable the French machines to break through into the British tunnels. The heads were buried in concrete, and the back-up equipment removed.

Below left: This probe drill was attached to the British service tunnel TBM. In October 1990, it was used to check the alignment of the British and French tunnels in advance of the breakthrough proper.

Below: Jubilant workers celebrate the final breakthrough. The French southern rail tunnel TBM broke into the British tunnel on 28 June 1991.

Before the breakthroughs, the cutting heads of British TBMs were buried in concrete. This was less expensive than extracting them from 18km (11 miles) of tunnel. The French TBMs were dismantled and taken back to Sangatte, where they were sold for scrap.

Above: In the control cabin of the British service tunnel TBM, radio confirmation that the probe has emerged into the French tunnel is eagerly awaited.

Right: Phillippe Cozette (right) and Graham Fagg were the tunnellers selected by ballot to make the service tunnel breakthrough.

THE COMING OF THE RAILWAY

The breakthrough was only the end of the first phase of construction, however. The narrow-gauge tracks of the construction railway had to be removed. Then the tunnels had to be cleared of debris and cleaned up. The French came up with an ingenious machine called the **diplodocus**, which combined these operations in one.

A concrete floor was then laid in the rail tunnels. This supports the standard gauge (1.435m/4ft 8.5in) railway track. Special concrete fixing blocks with rubber shock-absorbers were used to secure the track. In all, 334,000 blocks were needed for the 100km (60 miles) of track. Finally, work was completed on the lighting, pipework and emergency communication systems.

EARTHQUAKES

Britain and France are not normally associated with earthquakes, but these were one of the potential hazards that the Tunnel's designers took into account — even though the last known major earth tremor in the region took place in 1531! Nevertheless, the Tunnel has been designed so that it will not collapse, in the unlikely event of an earthquake.

Below: The giant track-laying machine at work in the UK undersea crossover cavern.

Left: The cooling pipes had to be welded before being bolted into position.

Below: The track was delivered in 180m (590ft sections). Some rails had to be cut to allow for the curvature of the tunnels.

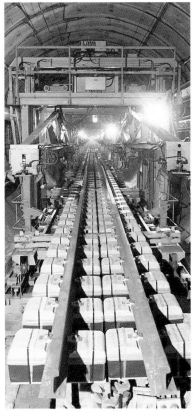

Above: A machine laying track in one of the British tunnels. The track came already attached to concrete fixing blocks, which were fitted with special shock-absorbing pads. It was then secured into a concrete track bed, to provide firm, smooth running for the exceptionally long and heavy trains.

BUILDING THE TERMINALS

While it was being built, the French terminal was the largest civil construction project in the world. Its 700 hectares (283 acres) make it Europe's largest land transport complex. Within its 18km (11 mile) perimeter fence, it has 50km (30 miles) of railway tracks and 50km (30 miles) of road. A huge shopping and leisure development called Le Cité de L'Europe is also being built nearby. It is due to be opened in spring 1995.

On the British side, because less space was available, the terminal site is more compact. Everything — rail and roads, terminal buildings, maintenance sheds and customs and passport facilities — had to be packed into a site of just 140 hectares (346 acres).

Above: Huge cranes were used to build the overbridges, which carry the cars and trucks on to and off the platforms at the terminals. In the background is the portal of the arrival loop tunnel.

Left: The concrete British portal was cast around a steel reinforcement cage.

Above: One of the massive vehicles used in the construction of the Calais Terminal.

Right: Track laying in progress at the Folkestone Terminal.

Below: One of the overbridges at the Calais Terminal under construction.

PRESERVING THE ENVIRONMENT

A huge engineering project like the construction of the Channel Tunnel is bound to have some influence on its immediate environment. However, Eurotunnel made a public promise to do everything possible to minimize the impact. Calais was virtually a "green field" site, so the French terminal could be built with little inconvenience to the local population. On the British side, detailed environmental surveys were carried out in the places that would be affected by the construction work.

One of the most important of these areas is a valley called Holywell to the east of the Folkestone Terminal. The three landward tunnels pass under the valley. A technique called "cut and cover" — which involves digging a trench that is covered over once the tunnel has been built — was used in building the tunnels at this point. The area has since been carefully restored to open meadow.

The site of Folkestone Terminal itself needed to be levelled. To avoid extensive excavations or truck movements, sand was dredged from the Goodwin Sands using a 5km (3 mile) pipeline. Three historic buildings on or near the Terminal site were carefully dismantled before being rebuilt elsewhere. Even a colony of great crested newts was found a new home.

Above: Tree planting being carried out at Dollands Moor. This is the location of British Rail's freight sidings, and is also where the exit route from the Folkestone Terminal joins the M20. The area surrounding the Terminal has been planted with half a million trees. This work was put in hand as early as possible, so that the Terminal would be well screened by the time it was officially opened.

Left: Recording the growth of plant life on the escarpment north of the Folkestone Terminal. This area, where rare orchids and butterflies can be found, has been officially designated a Site of Special Scientific Interest.

Opposite page: The coastal footpath at Shakespeare Cliff. The ventilation shaft on the right comes from a tunnel on the Dover to Folkestone railway line.

THE FINISHING TOUCHES

In order to blend in with their surroundings, the permanent buildings at the Folkestone Terminal have been restricted to four storeys in height. The whole area around the Terminal is being landscaped with half a million trees and shrubs. Earth embankments topped with acoustic screens help to reduce noise and ensure that the Terminal buildings are out of sight of neighbouring villages. The western end of the shuttle rail loop runs below ground, again to reduce noise levels. Lighting at the terminal has also been carefully designed so as not to "spill" outside.

Windbreaks were erected to protect the high-sided shuttles from strong gales. Tall fences, well above and below ground level, prevent animals from getting anywhere near the Tunnel. Electrical substations were built, to provide power for the catenary, which supplies electricity to the terminals as well as to the shuttles and trains in the tunnels. And signposts were put up, with directions for cars, coaches and trucks.

Above: The earth embankment at the landward end of Folkestone Terminal screens it from nearby villages.

Left: The motorway interchange at the Calais terminal is built round a huge artificial lake, making a spectacular entrance route.

Below: Homeward-bound Britons approach the Terminal at Calais.

Right: Adding Le Shuttle livery to a brand new shuttle locomotive.

THE TUNNEL'S ROLE

The Channel Tunnel is now an important and integral part of the European transport system. It provides a direct link between British and Continental road and rail networks. The Tunnel has greatly reduced the time British motorists take to get to their destinations on the Continent. It has also greatly speeded up journey times for cross-Channel truck drivers.

In addition, the Channel Tunnel connects Britain to the high-speed rail system that carries passengers between major European cities in times that rival those of air travel — and gets them straight to city centres rather than to airports well outside the city limits.

European rail companies are planning 23,000km (14,375 miles) of high-speed line. Some of this , such as the line linking the Tunnel to Paris, is already in use, while more is under construction. When the high-speed line is built connecting the Tunnel to London — possibly at the beginning of the next century — it will further reduce rail journey times (currently 3 hours from London to Paris and 3 hours 10 minutes from London to Brussels).

Up to 40 high-speed trains a day are planned, travelling at speeds of up to 300km/h (185mph). Thirty-eight Eurostar Trains have already been ordered. Each train will be air-conditioned, with 18 carriages and nearly 800 seats.

There are also plans to introduce night "hotel" trains with sleeper cars containing beds, toilets and showers, as well as carriages with reclining seats. These trains will each carry some 400 passengers, and run from Plymouth, Swansea and Glasgow to Amsterdam, Paris and Brussels, and from London to Amsterdam, Frankfurt and Dortmund.

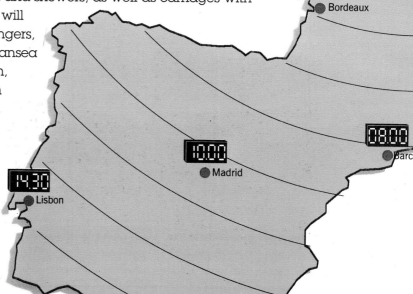

A map of Europe showing travel times to major European cities from the mid-point of the Tunnel in about the year 2010, when the plans for a European high-speed network have been fulfilled.

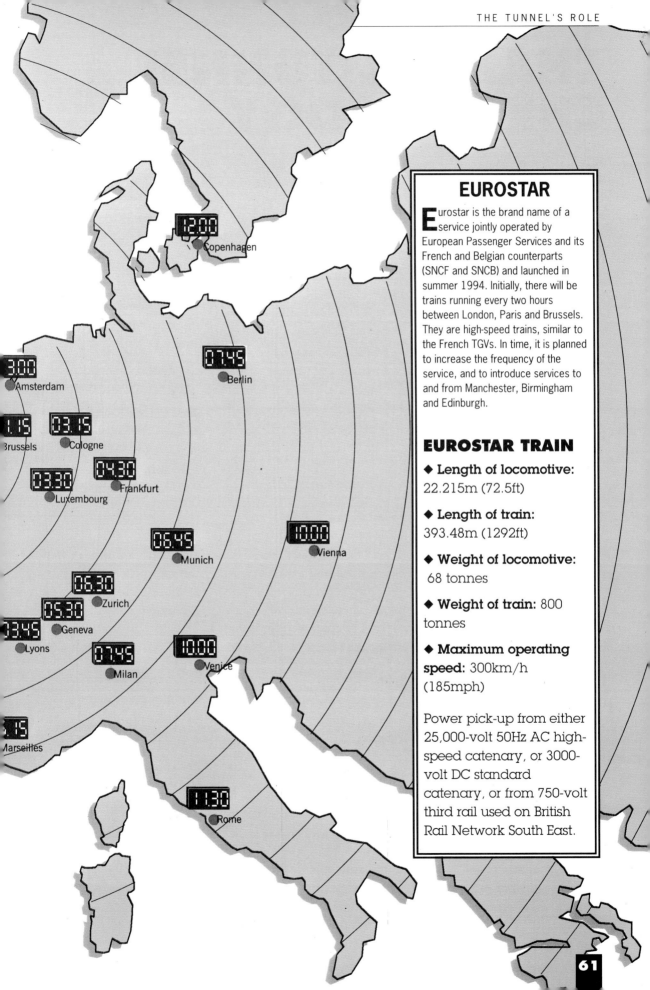

12.00 Copenhagen

07.45 Berlin

3.00 Amsterdam

1.15 Brussels

03.15 Cologne

02.30 Luxembourg

04.30 Frankfurt

06.45 Munich

10.00 Vienna

06.30 Zurich

05.30 Geneva

03.45 Lyons

07.45 Milan

10.00 Venice

15 Marseilles

11.30 Rome

EUROSTAR

Eurostar is the brand name of a service jointly operated by European Passenger Services and its French and Belgian counterparts (SNCF and SNCB) and launched in summer 1994. Initially, there will be trains running every two hours between London, Paris and Brussels. They are high-speed trains, similar to the French TGVs. In time, it is planned to increase the frequency of the service, and to introduce services to and from Manchester, Birmingham and Edinburgh.

EUROSTAR TRAIN

◆ **Length of locomotive:** 22.215m (72.5ft)

◆ **Length of train:** 393.48m (1292ft)

◆ **Weight of locomotive:** 68 tonnes

◆ **Weight of train:** 800 tonnes

◆ **Maximum operating speed:** 300km/h (185mph)

Power pick-up from either 25,000-volt 50Hz AC high-speed catenary, or 3000-volt DC standard catenary, or from 750-volt third rail used on British Rail Network South East.

CHANNEL TUNNEL CHRONOLOGY

1802
Albert Mathieu-Favier puts forward first documented scheme for Channel tunnel

1833
M.J.A. Thomé de Gamond makes first of several proposals for a tunnel

1867
Thomé de Gamond presents further scheme at Paris Great Exhibition

1872
Anglo-French Channel Tunnel Committee formed

1881
Tunnelling work starts at Shakespeare Cliff with Beaumont-English machine

1882
Work is stopped because of fear of invasion

1906
New scheme proposed (unsuccessfully) by Channel Tunnel Company

1922
Trial tunnel excavation starts in cliffs near Folkestone

1929
Royal Commission re-examines feasibility of tunnel, but Parliament votes against proposal

1974
Work on tunnel restarted, but halted the following year on economic grounds

1980
British House of Commons Select Committee begins new feasibility study for a privately-financed fixed Channel link

1981
British Prime Minister Margaret Thatcher and President Mitterrand of France announce new studies for fixed Channel link

1984
30 November Margaret Thatcher and President Mitterrand issue joint statement saying that "a fixed cross-Channel link would be in the mutual interests of both countries"

1985
The two governments invite submission of plans. Nine schemes are submitted

1986
20 January Scheme submitted by Channel Tunnel Group Ltd and France-Manche SA chosen
12 February Treaty of Canterbury signed, which enables both parliaments to pass the necessary laws

1987
Legislation passed by British and French parliaments
1 December British undersea service tunnel started

1988
28 February French undersea service tunnel started

1990
30 October Technical breakthrough by undersea service tunnel probe
1 December Undersea service tunnel breakthrough

1991
22 May Northern undersea rail tunnel breakthrough
28 June Southern undersea rail tunnel breakthrough

1992
14 December First shuttle locomotive delivered to Calais Terminal

1993
10 December System handed over to Eurotunnel by TML

1994
6 May Channel Tunnel officially opened by HM Elizabeth II and President Mitterrand

GLOSSARY

Access shaft A passage used during construction which led from ground level to the tunnel workings.

Bogie A frame containing wheels and axles, capable of rotating under a rail locomotive or item of **rolling stock**, to enable it to go round corners.

Catenary The overhead cable running through the rail tunnels that supplies power to the trains.

Chalk marl A soft but firm and **impervious** rock, ideal for tunnelling. The line of the tunnel follows a layer of chalk marl running beneath the Channel.

Crossover cavern A huge underground chamber, where trains can change from one tunnel to the other when part of a tunnel is closed down for maintenance or repairs. There are two crossover caverns, each located about one-third of the distance from the nearest portal.

"Diplodocus" A machine so named by French construction workers. It removed the temporary tracks of the construction railway at the same time as clearing the debris from the tunnels.

"Disneymobile" A special type of glass-roofed **manrider**, used for carrying visitors along the construction railway.

Electrical regeneration One of the shuttle locomotive's braking systems, in which the wheels power the motors instead of the other way round, thus generating electricity instead of using it.

Flange The raised edge of a wheel. Flanged wheels are used on railway locomotives and rolling stock to prevent them slipping off the tracks.

Flatbed A flat-topped railway wagon used for transporting supplies.

Grout A waterproof material used for sealing the joints between tiles and other surfacing materials. In the Tunnel, the lining segments were grouted with cement.

Impervious Waterproof. The term is used of rock, such as chalk marl, which does not allow water to seep through it. Pervious rocks, on the other hand, permit the passage of water.

Le Shuttle Eurotunnel's brand name for its cross-Channel service for passenger and freight vehicles.

Manrider A specially-designed train that took staff to their workplaces along the construction railway.

Mimic board A visual display in the rail control centre at Folkestone showing the position of all the traffic in the tunnels.

Muck wagon An item of **rolling stock** used on the construction railway to remove spoil from the tunnels.

Pantograph A device fixed to the roof of a locomotive that picks up electric power from the **catenary**.

Piston relief duct A tube, equipped with valves, that links the two rail tunnels. Its purpose is to equalize air pressure in the rail tunnels by allowing air that builds up in front of a train in one tunnel to flow into the other tunnel.

Rail tunnel One of the two tunnels used by shuttles and through-trains. (The term is used to distinguish the rail tunnels from the **service tunnel**).

Rake The name given to a group of carriages on the **shuttle**. Passenger-vehicle shuttles each consist of two rakes — one made up of twelve single-deck carriages and the other of twelve double-deck carriages. In addition, each rake has a loading carriage at one end and an unloading carriage at the other. Freight shuttles consist of two rakes of 14 carrier wagons each. Again, each rake is equipped with a loading and an unloading wagon.

Rolling stock The carriages and wagons that run on a railway (the term is also sometimes used to include locomotives).

Service tunnel The smaller central tunnel, running between the two rail tunnels, for use in routine maintenance.

Shuttle A train or other form of transport moving back and forth between two places. Eurotunnel shuttles run between the terminals at Folkestone and Calais. (The term "shuttle" comes from an old English word for a dart, and was originally used in weaving for a device that was shot to and fro between the threads.)

Slurry A semi-liquid paste. At Sangatte, spoil from the tunnel workings was mixed with water to form slurry that could be piped to the artificial lake at Fond Pignon.

Spoil Rock dug out of the tunnel workings.

STTS (Service Tunnel Transportation System) vehicle A specially-designed vehicle used in the **Service Tunnel** for maintenance and emergency purposes.

Supplementary ventilation system An emergency supply of fresh air for the tunnels, only used during maintenance or if there is a problem with the main ventilation system. The fans used in the system are reversible, so that they can be used either to blow air into the tunnels or remove smoke from them.

TBM (Tunnel Boring Machine) A specially-designed machine for digging the tunnel. In all, 11 TBMs were used to excavate the three tunnels.

TGV (*Train à grande vitesse*) A high-speed train, used on the French rail system. It can cruise at speeds of up to 300km/h (185mph) on specially-laid tracks, and its maximum speed is considerably higher — the record is 515km/h (322mph).

Train captain The official on **Le Shuttle** who is responsible for the safety of the train and its passengers.

VDU The abbreviation for a Visual Display Unit, on whose screen information is displayed electronically.

INDEX